BUILDING BLOCKS OF COMPUTER SCIENCE

SEMICONDUCTORS

Written by Echo Elise González

Illustrated by Graham Ross

WORLD BOOK

www.worldbook.com

Co-published by agreement between Shi Tu Hui and World Book, Inc.

Shi Tu Hui
Room 1807, Block 1,
#3 West Dawang Road
Chaoyang District, Beijing 100025
P.R. China

World Book, Inc.
180 North LaSalle Street
Suite 900
Chicago, Illinois 60601
USA

Library of Congress Control Number: 2025938322

Building Blocks of Computer Science
ISBN: 978-0-7166-6706-3 (set, hard cover)

Semiconductors
ISBN: 978-0-7166-6715-5 (hard cover)

Also available as:
ISBN: 978-0-7166-6725-4 (soft cover)
ISBN: 978-0-7166-6735-3 (e-book)

WORLD BOOK STAFF

Editorial

Vice President
Tom Evans

Senior Manager, New Content
Jeff De La Rosa

Manager, New Product Development
Nicholas Kilzer

Associate Manager, New Content
William D. Adams

Content Creator
Elizabeth Huyck

Proofreader
Nathalie Strassheim

Graphics and Design

Senior Visual Communications Designer
Melanie Bender

Acknowledgments
Writer: Echo Elise González
Illustrator: Graham Ross/The Bright Agency
Series reviewed by George K. Thiruvathukal (Loyola University Chicago); Peter Jang (Actualize Coding Bootcamp)

TABLE OF CONTENTS

There is a glossary on page 39. Terms defined in the glossary are in type **that looks like this** on their first appearance.

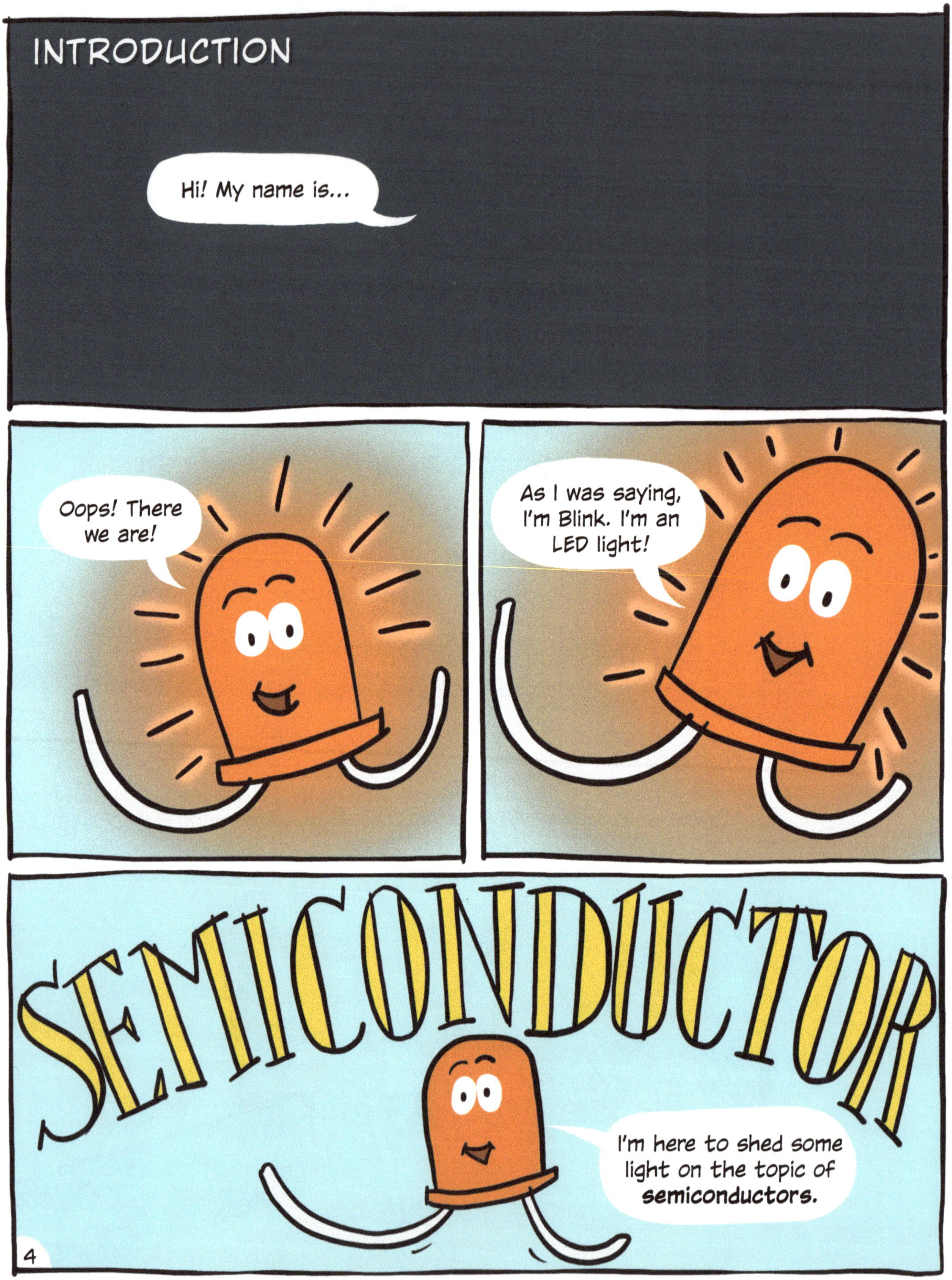
INTRODUCTION
Hi! My name is...
Oops! There we are!
As I was saying, I'm Blink. I'm an LED light!
SEMICONDUCTOR
I'm here to shed some light on the topic of **semiconductors.**

What are they? How do they work? Why are they important?
It's all about controlling electricity.
Some materials, such as many metals, let electricity flow easily through them.
Other materials, such as glass and rubber, block electricity.
A semiconductor can do both. It lets electricity flow only under certain conditions.
Semiconductors are used to make the **computer chips** that power most digital devices we use today. Why? Let's find out!
CLICK

INSULATORS, CONDUCTORS, AND SEMICONDUCTORS
Electricity is **electrons** flowing. You can think of it a bit like water.
Imagine you want to water your flowers.
A hose will let water through easily. A little too easily!
CONDUCTOR
A material that lets electricity flow through it is called a **conductor.**
INSULATOR
An **insulator** is a material that electricity cannot pass through.

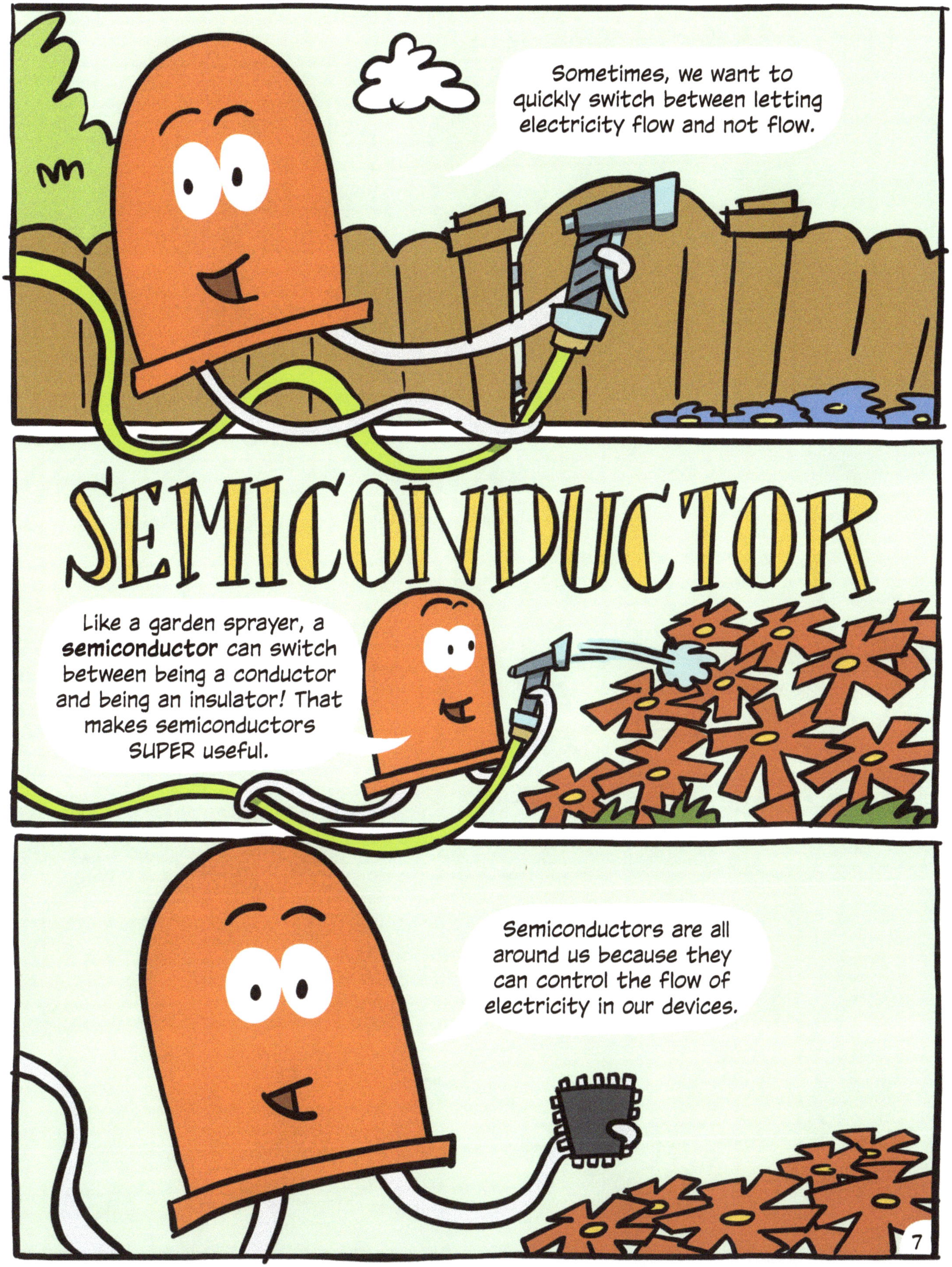

Sometimes, we want to quickly switch between letting electricity flow and not flow.
SEMICONDUCTOR
Like a garden sprayer, a **semiconductor** can switch between being a conductor and being an insulator! That makes semiconductors SUPER useful.
Semiconductors are all around us because they can control the flow of electricity in our devices.

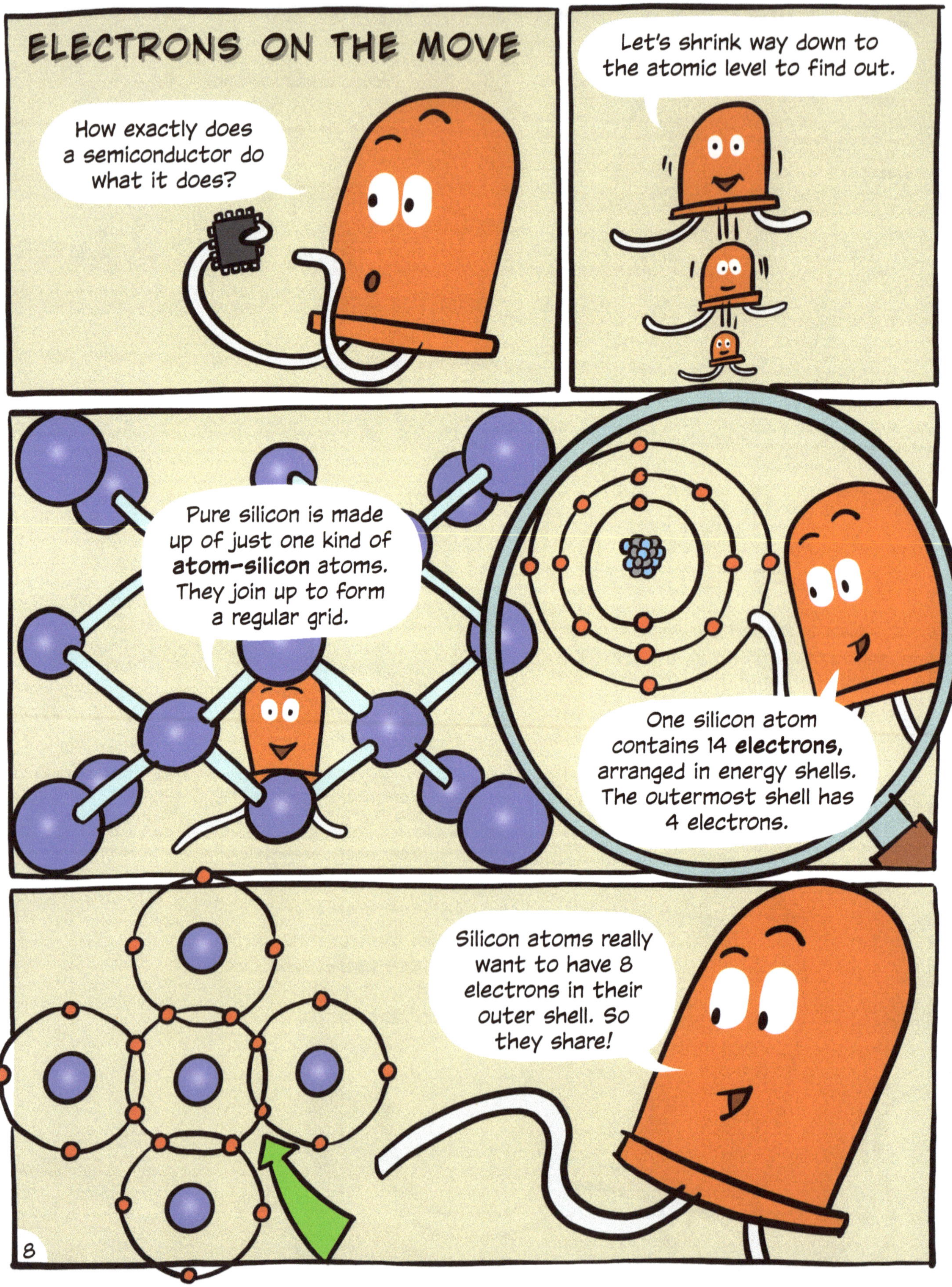
ELECTRONS ON THE MOVE
How exactly does a semiconductor do what it does?
Let's shrink way down to the atomic level to find out.
Pure silicon is made up of just one kind of atom–silicon atoms. They join up to form a regular grid.
One silicon atom contains 14 electrons, arranged in energy shells. The outermost shell has 4 electrons.
Silicon atoms really want to have 8 electrons in their outer shell. So they share!
8

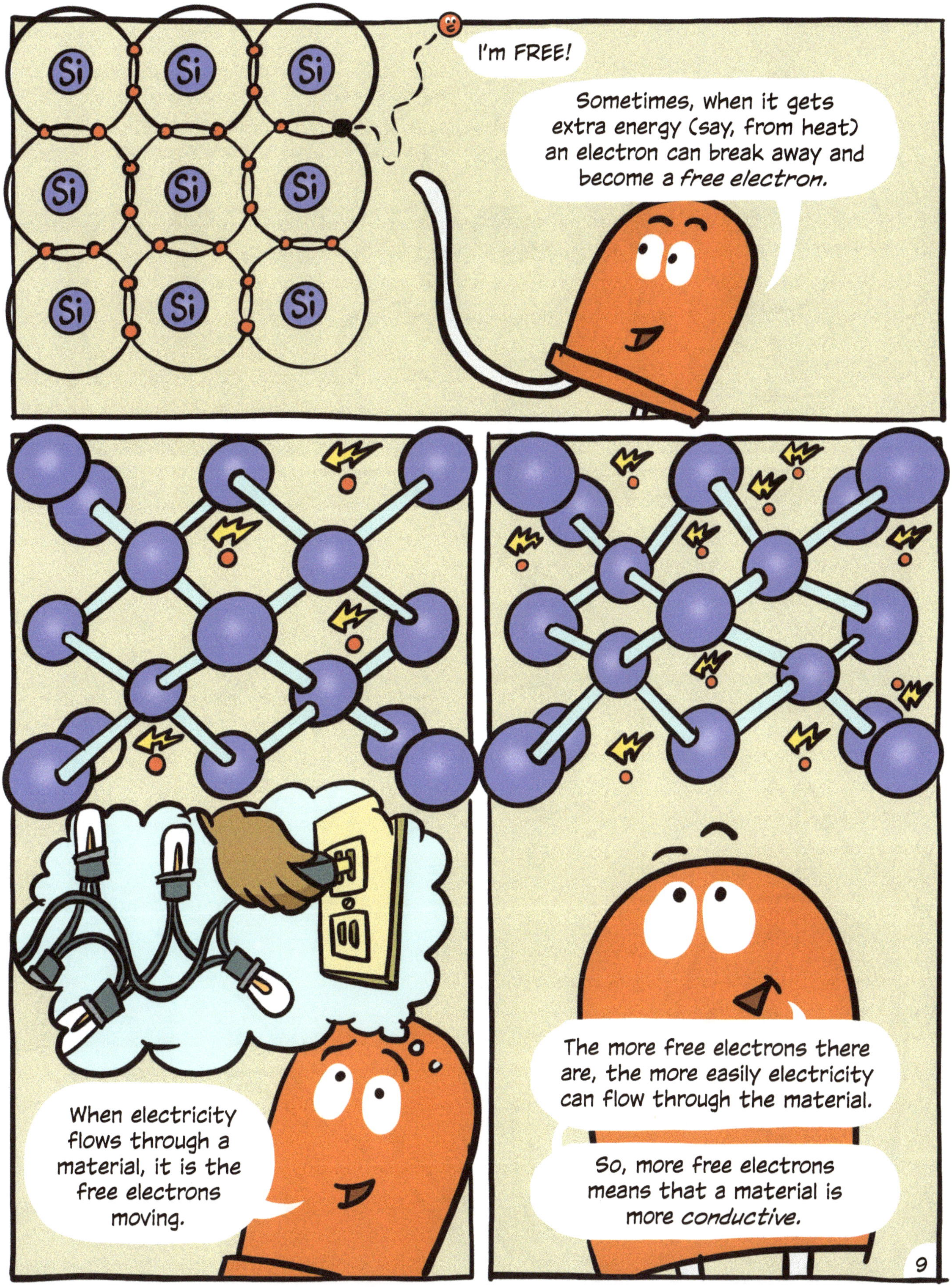
Si
I'm FREE!
Sometimes, when it gets extra energy (say, from heat) an electron can break away and become a *free electron.*
When electricity flows through a material, it is the free electrons moving.
The more free electrons there are, the more easily electricity can flow through the material.
So, more free electrons means that a material is more *conductive.*

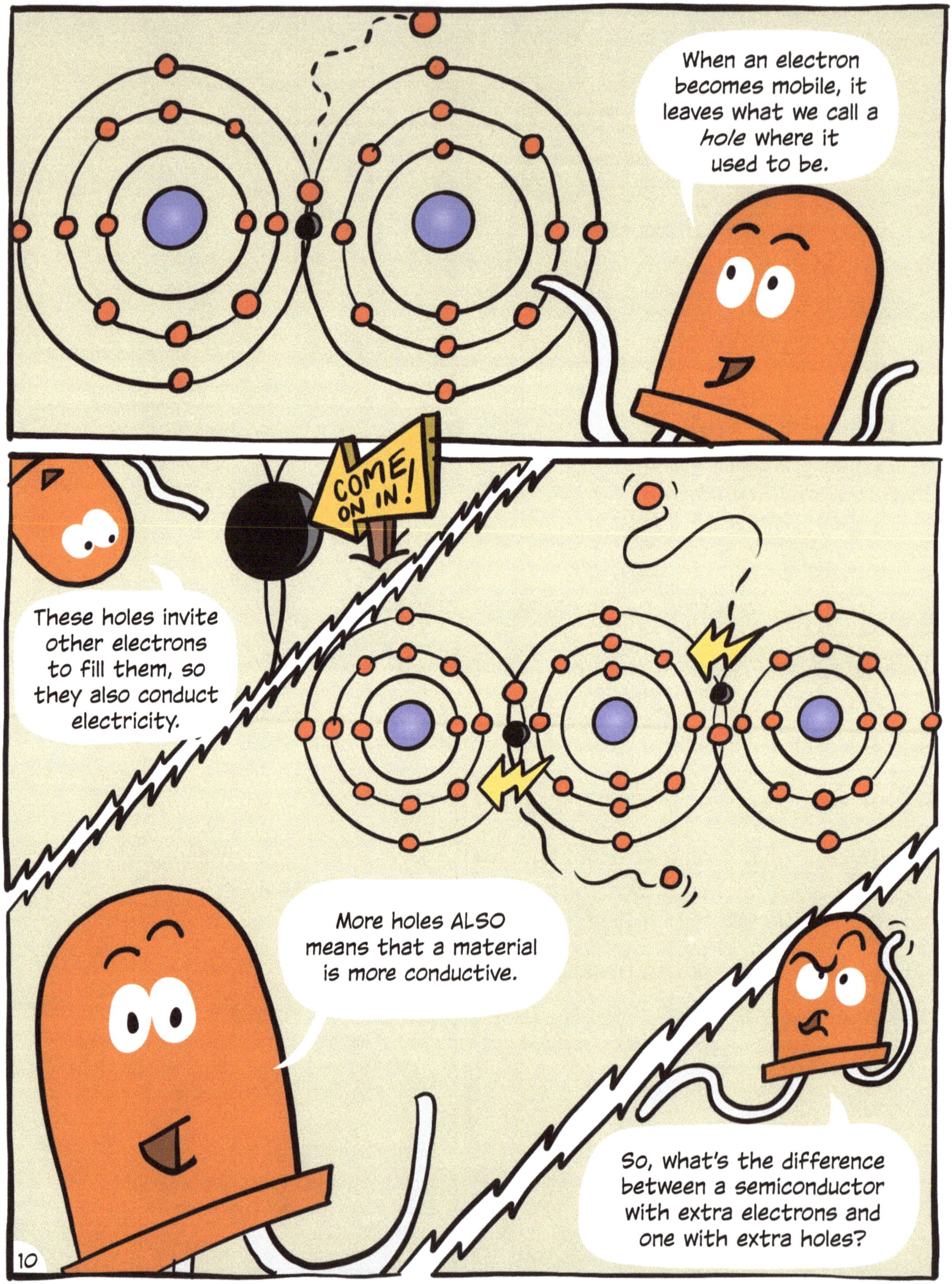
When an electron becomes mobile, it leaves what we call a *hole* where it used to be.
COME ON IN!
These holes invite other electrons to fill them, so they also conduct electricity.
More holes ALSO means that a material is more conductive.
So, what's the difference between a semiconductor with extra electrons and one with extra holes?

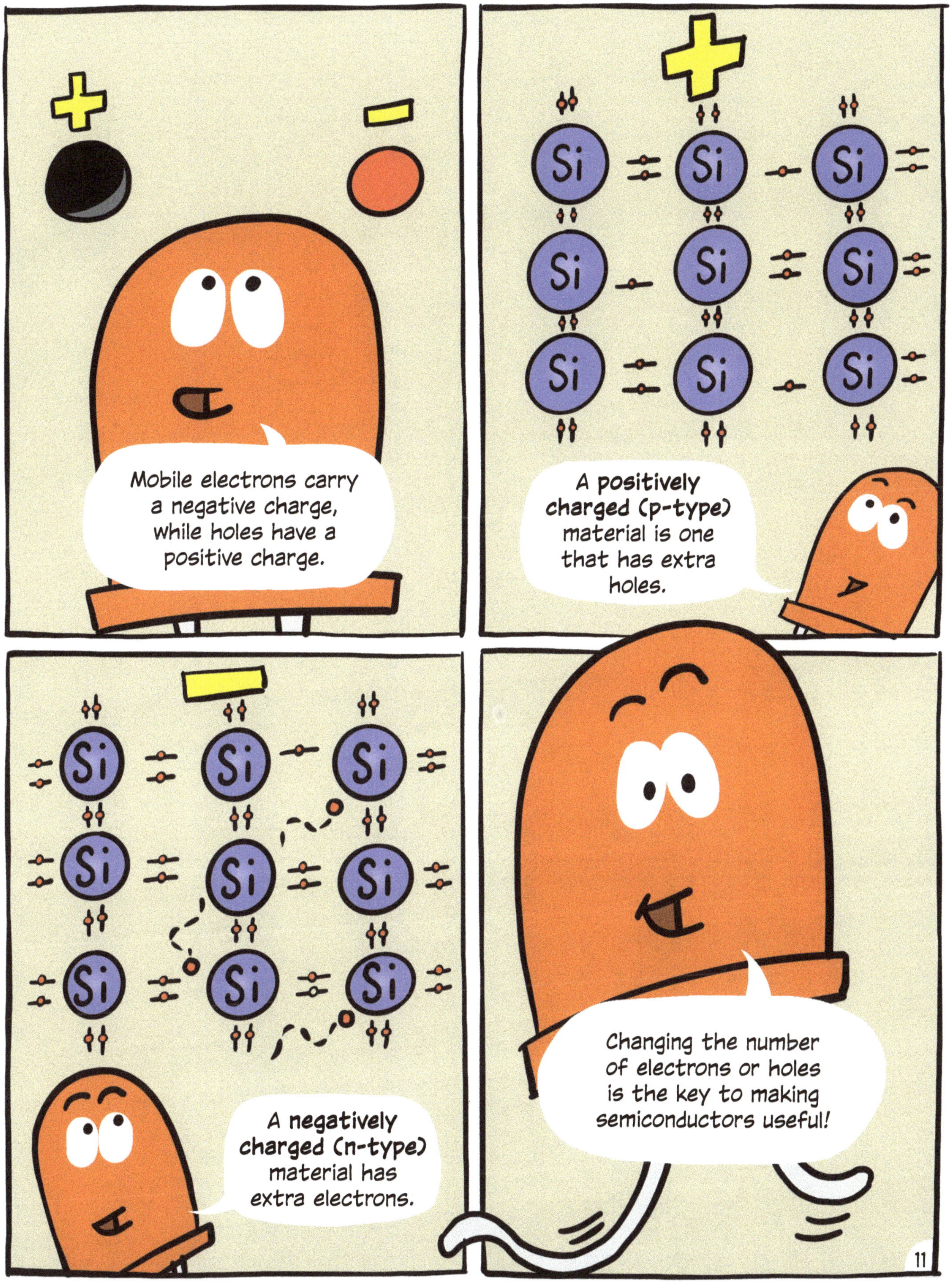
Mobile electrons carry a negative charge, while holes have a positive charge.
A positively charged (p-type) material is one that has extra holes.
Si
A negatively charged (n-type) material has extra electrons.
Changing the number of electrons or holes is the key to making semiconductors useful!

By itself, pure silicon is not a very good **conductor.**
To turn silicon into a true semiconductor...
LAB
You need to add a little seasoning–a few atoms of another **element.**
That's called **doping** the silicon.

Si
Si
Si
Free Electron
Si
P
Si
Si
Si
Si
Swapping in atoms with five electrons in their outer shell (such as phosphorus) adds free electrons.

B
Si
Si
Hole
Hole
Si
Si
B
Hole
B
Si
Si
Adding atoms with three electrons (such as boron) adds more holes.
By controlling exactly how many atoms are added, and what kind, engineers can create semiconductors with different properties.

To see how p and n doping changes silicon, let's play musical chairs!
OR
AND
If we have the same number of players and chairs, that's like plain silicon.
Hmm... Because the number of chairs and players is the same, there isn't much action.
OR
AND
We need an extra player.
Great idea!
This is more like it!
AND
OR
With one extra player, someone is always moving. The electricity is flowing!
AND
OR

This puzzle has no empty spaces. So nothing moves! That's also like regular silicon.
Let's see if we can get things moving.
POP
Much better! With a hole, tiles can flow.
In p-doped silicon, the holes pull electrons to keep them moving.
Whether you have extra electrons or extra holes, current flows!

DEPENDABLE DIODES
We're going to the movies!
OR
AND
We've got our tickets so we're entering through the front doors to find our seats.
ENTER
AND
OR
ENTER
Great flick! Looks like we can't exit through the same door we came in.
OR
EXIT
AND
OR
We have to go out the EXIT door.
You know, this reminds me of how a **diode** works!

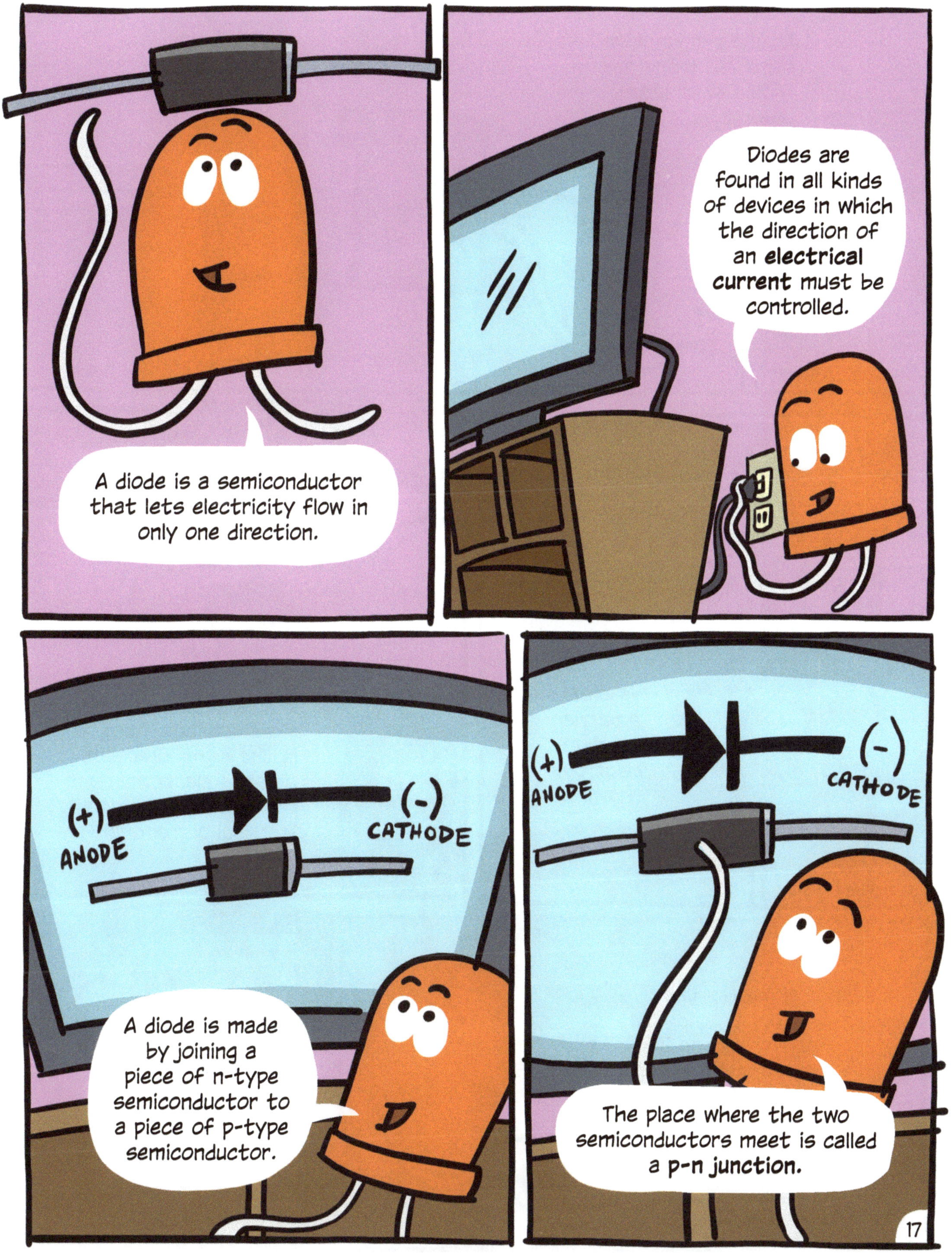
A diode is a semiconductor that lets electricity flow in only one direction.
Diodes are found in all kinds of devices in which the direction of an **electrical current** must be controlled.
(+)
ANODE
(-)
CATHODE
A diode is made by joining a piece of n-type semiconductor to a piece of p-type semiconductor.
(+)
ANODE
(-)
CATHODE
The place where the two semiconductors meet is called a **p-n junction.**

At the p-n junction, some electrons and holes swap sides.
P-TYPE
N-TYPE
This creates a barrier to more electrons flowing.
HOLES
ELECTRONS
If we connect a battery in the right direction, it pushes electrons past the barrier, allowing the electrical current to flow.
If we flip the battery around, the barrier widens and stops the current from flowing.
BARRIER

I'm a diode too! **LED** stands for *light-emitting diode.*
LED's are made of semiconductors doped with atoms that emit light when their electrons get excited.
That's what makes me shine!
The color of the light depends on what elements are added.
Aluminum indium gallium phosphide makes me glow orange!
LED's are used in all kinds of screens. Computers, tablets, phones...
...traffic lights, bike lights, holiday lights—and energy-saving light bulbs!

TERRIFIC TRANSISTORS

OFF
ON
This can turn the sandwich into a switch, letting electricity through or not depending on whether the gate part has an electric charge or not.
A small electric current can let a much larger one flow! Doping can control how strong the charge has to be to open the gate.
OFF
GATE
n
p
n
In an actual transistor, the gate is p- and n-doped semiconductors. With no charge (or not enough), no electricity flows.
ON
n
p
n
When electricity flows to the gate, it opens a path. And electricity flows!

SO MUCH SILICON
In electronics, silicon is the most commonly used semiconductor.
Semiconductors can also be made from germanium, carbon, and many metal compounds.
But silicon is the star!
SILICONE
OTHER ELEMENTS
Silicon makes up about 28 percent of Earth's crust. It's the main ingredient in beach sand and many rocks.
FAB
It is the second most abundant element on Earth, after oxygen. So, it's easy to get.

Silicon's abundance and chemical structure makes it especially useful for electronics.
CLUNK
Engineers first purify the silicon.
Then they grow it into big crystal cylinders.
The cylinders are sliced into very thin wafers...
...which are used to make **microchips.**

SEMICONDUCTORS IN MICROCHIPS

Hey! What's the big idea?
Who keeps turning out the lights?
A transistor acts like a teeny tiny version of a light switch.
When the switch is on, it lets a current of electricity through. When it's off, no current flows.
What does this have to do with computers?
Computer programs run on a code called **binary.** Binary works with just two numbers, 0 and 1—or "off" and "on."
Transistors are the physical parts that flip off and on to make all those 0's and 1's, millions of times a second.

Microchip engineers combine transistors to make **logic gates** in a microchip.
Logic gates are the parts of a microchip that do math and logic functions.
We can program what kind of decision the chip will make by arranging transistors into the right kind of logic gate.

Let's say we connect two transistors in a row, a bit like this.

CLICK
In this arrangement, electricity can flow only if BOTH transistors are switched to "on."
CLICK
AND
This kind of logic gate is called an *AND* gate.
CLICK
In an AND gate, if one of the transistors is "off," then the electricity cannot pass.
AND

Let's look at another logic gate: the OR gate!
But in this gate, the transistors are parallel instead of being connected in a row.
OR
CLICK
OR
Like the AND gate, the OR gate also has two connected transistors.
If this transistor OR that transistor is on, then the output is "on."
CLICK
OR
So, even if one of the transistors is "off," the output can still be "on."
In order for the output to be "off," both transistors have to be "off."
OR
OR

Logic gates like AND and OR are the building blocks of microchips.
OR
By combining series of many logic gates, engineers create chips that can do math, store data, and run programs.
3
00000011
3
00000011
+
6
00000110
A B CI
0 0 1
Adder (C)
0 1
CO Sum
A B CI
1 1 1
Adder (B)
1 1
CO Sum
A B CI
1 1 0
Adder (A)
1 0
CO Sum
4
2
1
Computer chips hold billions of microscopic logic gates made of semiconductors.
OR
Without logic gates, computers wouldn't be able to think!
And without transistors, there would be no logic gates.
OR
AND
That's why semiconductors are so important!

SEMICONDUCTOR SUPERPOWERS
Semiconductors are used to make the computer chips in our computers, smartphones, and tablets...
And many other devices! Microwave ovens, washing machines, smart speakers, thermostats, and video game consoles all use semiconductors.
The sensors in my car are made with semiconductors.
So are the traffic lights.

Radar and airplane navigation tools use semiconductors.
HOSPITAL
So do satellites and cell towers.
X-RAY DEPARTMENT
X-ray machines and CT scanners could not function without semiconductors.
Engineers are constantly working to make even smaller transistors, to help **artificial intelligence (AI)** process information quickly.
Semiconductors make the digital world possible! Thanks, silicon.

TIMELINE

1833

British chemist Michael Faraday observed semiconducting, noting that silver sulfide conducted electricity better when heated, unlike most metals.

1883

American Charles Fritts constructed the first working solar cell out of semiconductor selenium and a thin layer of gold.

1940's

Semiconductors were developed for use in radar detectors during World War II.

1947

Americans William Shockley, Walter Brattain, and John Bardeen invented the transistor, a fast semiconductor-based electronic switch. Engineer John Pierce, who wrote science fiction on the side, came up with the name "transistor."

1957

The Sony company marketed one of the first portable transistor radios. It became a best seller and ushered in a new era of personal electronics.

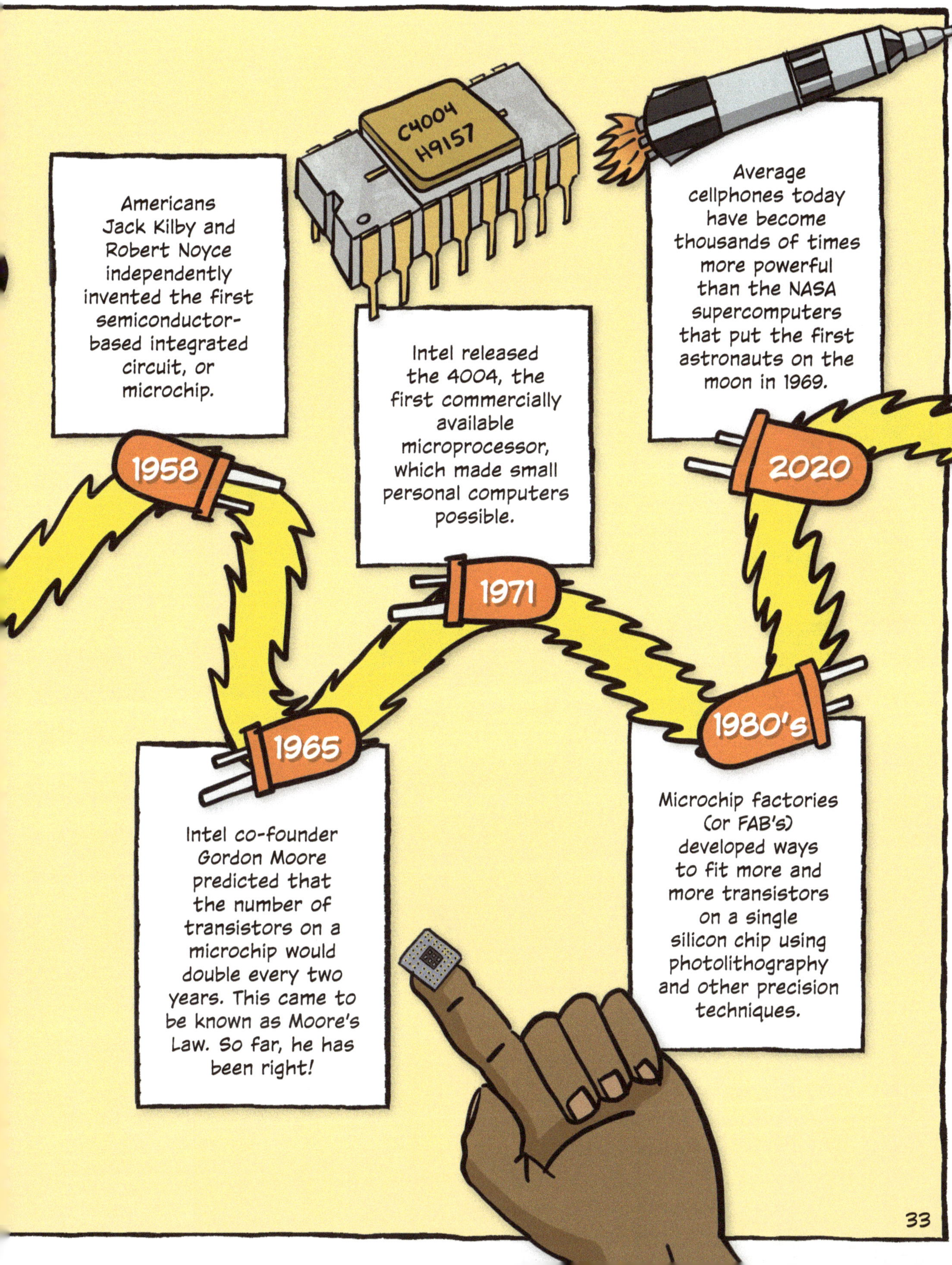
1958
Americans Jack Kilby and Robert Noyce independently invented the first semiconductor-based integrated circuit, or microchip.
1965
Intel co-founder Gordon Moore predicted that the number of transistors on a microchip would double every two years. This came to be known as Moore's Law. So far, he has been right!
C4004
H9157
1971
Intel released the 4004, the first commercially available microprocessor, which made small personal computers possible.
1980's
Microchip factories (or FAB's) developed ways to fit more and more transistors on a single silicon chip using photolithography and other precision techniques.
2020
Average cellphones today have become thousands of times more powerful than the NASA supercomputers that put the first astronauts on the moon in 1969.

WHO'S WHO:
JACK KILBY AND ROBERT NOYCE
Microchips are such a great idea...
Jack Kilby
Robert Noyce
They were invented twice!
Computers run on streams of binary digits: 0 or 1, off or on.
To make those binary digits, you need millions of switches that can turn off and on super fast.
ENIAC, one of the first working computers, used 18,000 vacuum tubes, similar to light bulbs, as binary switches.
The tubes were big, fragile, and often broke. They also used a huge amount of power!
In 1947, telephone engineers trying to improve on vacuum tubes discovered a way to make a switch called a transistor out of semiconductors.
Small! No moving parts! Uses less power!
Though they still had to be wired together and attached to a board.

I got into radios when I was a kid. After building radios for the Army in World War II, I got a job designing small electronic circuits for Texas Instruments.
I studied physics at MIT and later worked for William Shockley, one of the inventors of the transistor. He was smart, but hard to get along with.
?
My first summer at my new job, everyone else went on vacation. So I sat around the lab and thought about ways to make smaller circuits.
Fact File
Name: Jack St. Clair Kilby
Born: 1923, in Jefferson City, Missouri, USA
Occupation: Electrical engineer, inventor
Claim to fame: Co-inventor of the microchip, 1958. Also invented hand-held calculators.
Some of my co-workers and I left Shockley's lab and started our own company. In 1958, I was thinking about how to shrink circuits, too.
Fact File
Name: Robert Norton Noyce
Born: 1927, in Burlington, Iowa, USA
Occupation: Physicist, electrical engineer, inventor
Claim to fame: Co-inventor of the microchip, 1958. Co-founded Intel company.

Semiconductors were good for making tiny transistors, diodes, and capacitors. But no matter how small we could make the parts, we still had to wire them together. That limited how small devices could get. I was tinkering with an idea for grids when...I had it!
We had been experimenting with laying wires directly on a single piece of silicon to connected doped transistor areas. It occurred to me that we could also make capacitors and diodes by doping areas of silicon. And then I had it!
A GREAT IDEA!
Why not make ALL the components of a computer circuit by doping different spots on a SINGLE square of silicon! Lay on thin metal channels to connect them, and you'd have a whole circuit on a single base. You could put loads of transistors very close together.
Behold, the Integrated Circuit! Or, computer chip.
That's me!

We had the same idea!
72 14 66 16 76 78 42 N
26 17 27 16 30 27 23 29 22 11 13 N P
We each filed patents around the same time. I filed first, but Noyce's was more worked out, so it was approved first.
We agreed to share the credit.
My company asked me to find a use for the new chips, so I invented the pocket calculator.
intel
With some friends, I started a new company, called Intel, to make chips for computers.
All those circuits are built up in microscopic layers in high-tech chip factories.
Integrated circuits have continued to get smaller and smaller. Today, millions fit on a single chip!
Coolest idea ever!

CAN YOU BELIEVE IT?!

The first semiconductor transistors were made of

germanium, not silicon.

Silicon later won out because it is cheap, abundant, and more heat-resistant.

The first transistor, made in 1947, was the size of a shoebox. Transistors have continued to shrink, until now some are

smaller than bacteria.

Silicon Valley

is a nickname for an area around Sunnyvale, California, where many high-tech companies are based. It takes its name from the semiconductor used to make microchips.

WORDS TO KNOW

artificial intelligence (AI) a kind of computer program that uses data and logic to solve complex problems.

atom the basic building blocks of matter. Atoms are made up of neutrons, protons, and electrons.

binary digit a 0 or 1. These two digits make up machine language.

circuit a path along which electricity can flow.

computer chip a small chip of silicon holding tiny electronic circuits that perform logic functions.

conductor a material that electricity can flow through easily.

diode a semiconductor switch that lets electricity flow in only one direction.

doping adding atoms of other elements to a pure substance (such as silicon) to change its electrical properties.

electric current flowing electricity.

electron part of an atom, extremely small elementary particle with a negative charge. When electrons move, they generate electricity and an electromagnetic field.

element a pure substance made of only one kind of atom.

FAB short for "fabrication facility," a high-tech factory where semiconductors and computer chips are made.

insulator a material that electricity cannot flow through.

integrated circuit a device in which all the components of an electronic circuit are created on a single semiconductor base.

LED light-emitting diode. A one-way semiconductor switch doped with material that emits light when it is electrified.

logic gate transistors that process electrical signals in a way that performs a logic function.

microchip or computer chip a silicon chip containing many integrated circuits that perform logic functions and store data.

n-type semiconductor a semiconductor doped to have fewer free electrons and an overall positive charge.

p-n junction the boundary between p- and n-doped semiconductors

p-type semiconductor a semiconductor doped to have extra free electrons and an overall negative charge.

semiconductor a material that can act as either a conductor or an insulator, allowing electricity to flow through it only under certain conditions.

silicon a metallic element that is a common component of rocks and sand. Silicon is the most widely used semiconductor for computer chips and electronics.

transistor an electrical switch composed of semiconductor layers with no moving parts.

INDEX

www.ingramcontent.com/pod-product-compliance
Lightning Source LLC
Chambersburg PA
CBHW040235041126
54327CB00025B/2447
9780716667254